中国
国家公园

国家公园管理局 \ 编

（试点篇）

NATIONAL
PARKS
OF CHINA

(Pilot chapter)

中国林业出版社

黄河

长江

NATIONAL PARKS OF CHINA
Pilot chapter

中国国家公园试点篇

前言

自然，具有至诚至善至美的意味；自然，充满着生命的律动。人类从荒野走出，依存于这个星球上的森林、草原、山岭、河流、海洋繁衍生息。人类和地球上的万千生物共生共荣。

我国幅员辽阔、地貌复杂，丰富多样的生态系统孕育了无数的生命，使我国成为全球生物多样性最丰富的国家之一。

多年来，随着经济建设发展的脚步，生态保护事业更加蓬勃，为了使珍贵自然遗产得到更好的保护，我国正在构建以国家公园为主体、自然保护区为基础、各类自然公园为补充的自然保护地体系。习近平总书记指出："中国实行国家公园体制，目的是保持自然生态系统的原真性和完整性，保护生物多样性，保护生态安全屏障，给子孙后代留下珍贵的自然资产。这是中国推进自然生态保护、建设美丽中国、促进人与自然和谐共生的一项重要举措。"

草木蔓发

春山可望

2013年11月党的十八届三中全会首次提出建立国家公园体制，2015年国家发展改革委联合国家林业局等13部委印发《关于印发建立国家公园体制试点方案的通知》，2017年中央印发《建立国家公园体制总体方案》，2019年中央印发《关于建立以国家公园为主体的自然保护地体系的指导意见》，国家公园体制建设大幕徐徐拉开……

敬畏自然、尊重自然、保护自然是人类与自然和谐共生的前提和道德基础，正确处理好人与自然、保护与发展的关系，让当代人享受到天蓝地绿水净、鸟语花香的美好家园，为子孙后代留下珍贵的自然资产，是我们建立国家公园体制的初心和使命。

草木蔓发，春山可望！国家公园这颗美丽中国王冠上的明珠，必将展现出生态文明建设新时代的中国自信。我们将紧紧围绕在以习近平同志为核心的党中央周围，坚定不移地按照习近平新时代生态文明思想，推进自然生态保护，建设美丽中国！

FOREWORD

Nature implies sincerity, perfection and beauty. Nature is full of the rhythm of life. Human beings come out of the wilderness and depend on forests, grasslands, mountains, rivers and oceans on this planet to thrive. Human beings and thousands of living things on the earth are symbiotic and co-prosperous.

China has a vast territory and complex geomorphology. The abundant and diversified ecosystems have given birth to countless lives, making China one of the countries with the richest biodiversity in the world.

Thanks to the rapid economic development over the years, the cause of ecological improvement has become more vigorous. To better protect the precious natural heritages, China is building a national park-centered protected area system based on nature reserves and supplemented by natural parks. General Secretary Xi Jinping pointed out: " China's implementation of the national park system is targeted at maintaining the authenticity and integrity of the natural ecosystem, conserving the biodiversity, protecting ecological security barriers, and leaving precious natural assets to future generations. This is an important measure for China to promote ecological improvement, build a beautiful China and promote harmonious symbiosis between man and nature. "

Since the third Plenary session of the 18th CPC Central Committee first proposed a national park system in November 2013, the Pilot program for the development of the national park system has been gradually rolled out with the introduction of such policies as the Circular on Releasing the Pilot Program for the Establishment of a National Park System released in 2015 jointly

by the National Development and Reform Commission, the State Forestry Administration and other 13 ministries, the General Plan for the Development of a National Park System released in 2017 by the central government and the Guidelines on the Establishment of a National Park-centered Protected Area System released in 2019 by the central government.

Reverence for nature, respect for nature and nature protection are the premise and moral underpinning of a harmonious symbiosis between man and nature. Our original intention and mission to establish a national park system are to handle properly the relationship between man and nature, protection and development so that the current generation can enjoy a beautiful homeland with a clear blue sky, green landscape, and lucid waters, singing birds and flowers giving forth their fragrance, and to leave precious natural heritage to our future generations.

If grass and trees are sprouting up, can the lush mountains in spring be far behind? The national park, like a pearl in the beautiful Chinese crown, will certainly mirror the confidence of China in the new era of ecological civilization.

We will firmly support the CPC Central Committee with Comrade Xi Jinping at the core, and unswervingly promote ecological progress and build a beautiful China under Xi Jinping's thinking on ecological civilization in the new era.

目 录
CONTENTS

前 言 007	
FOREWORD 011	
概 述 017	
OVERVIEW 019	
试点篇 021	
PILOT CHAPTER	

022 **三江源**
037 国家公园
Three-River-Source National Park

038 **东北虎豹**
049 国家公园
Northeast China Tiger and Leopard National Park

104 **武夷山**
121 国家公园
Wuyishan National Park

122 **神农架**
141 国家公园
Shennongjia National Park

050 **大熊猫**
073 国家公园
Giant Panda National Park

074 **祁连山**
087 国家公园
Qilian Mountain National Park

088 **海南热带雨林**
103 国家公园
Hainan Tropical Rainforest National Park

大事记
CHRONICLE OF EVENTS
183-192

结语
CONCLUDING REMARKS
194-195

142 **普达措**
157 国家公园
Potatso National Park

158 **钱江源**
167 国家公园
Qianjiangyuan National Park

168 **南山**
181 国家公园
Nanshan National Park

概述

党的十八届三中全会首次提出建立国家公园体制，开启了我国自然资源保护的新篇章。2018年，新组建的国家林业和草原局（国家公园管理局）统一管理各类自然保护地，探索解决长期困扰自然保护地建设的多头管理、交叉重叠、碎片化保护等问题，这是推动自然保护地管理体制改革的重大突破，具有里程碑的重要意义。

国家公园是以保护具有国家代表性的自然生态系统为主要目的，是我国自然生态系统中最重要、自然景观最独特、自然遗产最精华、生物多样性最富集的部分。目前，全国有三江源、东北虎豹、大熊猫、祁连山、海南热带雨林、武夷山、神农架、普达措、钱江源、南山等10个国家公园体制试点，涉及12个省，总面积超过22万平方公里，约占国土陆域面积的2.3%。

《中国国家公园（试点篇）》全方位展现各试点区的基本情况、大美风光、濒危保护物种、丰富的生物多样性等内容。开启这本画册，就如同走进了国家公园的奇妙世界。这里山川秀美，鸟兽灵动，这里有故事，也有传奇，这里有许许多多的意外和惊喜。中国国家公园体制试点工作如同开往春天的列车，载着希望，奔向美好的明天！

OVERVIEW

The Third Plenary Session of the 18th CPC Central Committee proposed for the first time a national park system, opening a new chapter in the protection of natural resources in China. In 2018, the newly established National Forestry and Grassland Administration (National Park Administration) administers all kinds of protected areas, exploring and solving the problems of multiple leadership, overlapping and fragmented protection that have long plagued the development of protected areas. It is a break through in promoting the reform of the management system of protected areas and is of great significance.

The national park is aimed at protecting the representative national natural ecosystem. It is the most important part of China's natural ecosystem, the most unique natural landscape, the most essential natural heritages and the richest biodiversity. At present, there are 10 Pilot projects for the national park system in China, including the Northeast Three-River-Source, China Tiger and Leopard, Giant Panda, Qilian Mountain, Hainan Tropical Rainforest, Wuyishan, Shennongjia, Potatso, Qianjiangyuan and Nanshan, involving 12 provinces, covering a total area of more than 220,000 km^2, accounting for about 2.3% of the land area.

National Parks of China (Pilot trials) illustrates the basic situation, great beauty, endangered species and rich biodiversity of each Pilot area. Opening this album is like walking into the wonderful world of a national park. There are beautiful mountains and rivers, smart birds and animals, stories and legends, and there are many surprises. The Pilot program of China's national park system seems like a train to the spring, carrying the hope of a better future.

NATIONAL PARKS OF CHINA

中国国家公园

试点篇

PILOT CHAPTER

2015年12月，开始三江源国家公园体制试点。试点区位于青海省，包括长江源、黄河源、澜沧江源三个园区，平均海拔4713.62米，总面积为12.31万平方公里，涉及果洛州玛多县，玉树州治多县、曲麻莱县、杂多县4个县和可可西里自然保护区。这里是长江、黄河、澜沧江三条江河的发源地，素有"亚洲水塔"之称，是我国乃至亚洲重要生态安全屏障，拥有世界上高海拔地区独有的大面积湿地生态系统。依托三江源的生态资源，这里成为青藏高原特有的，以藏羚羊、野牦牛、藏野驴、藏原羚等为主的珍稀濒危动物的重要栖息与繁殖地，是世界范围内特有的高寒生物自然种质资源库。

The Pilot project of the Three-River-Source National Park was launched In December 2015. The Pilot area is located in Qinghai Province, including three parks, i.e. the source of the Yangtze River, the source of the Yellow River and the source of the Lancang River, with an average elevation of 4713.62 meters and a total area of 123100 km^2, involving Maduo County of Guoluo. Zhiduo County of Yushu Prefecture, Qumalai County, Zaduo County and Hoh Xil Nature Reserve. As the birthplace of the Yangtze River, the Yellow River and the Lancang River, known as the "Asia Water Tower", it is an important ecological security barrier in China and even Asia and has an endemic and vast area of wetland ecosystem in high-altitude areas of the world. Relying on the ecological resources of Three-River-Source, it has become an important habitat and breeding ground for rare and endangered animals endemic to the Qinghai-Tibet Plateau, such as Tibetan antelope, wild yak, Tibetan wild ass and Tibetan gazelle . It is an endemic alpine germplasm bank in the world.

三江源国家公园
Three-River-Source National Park

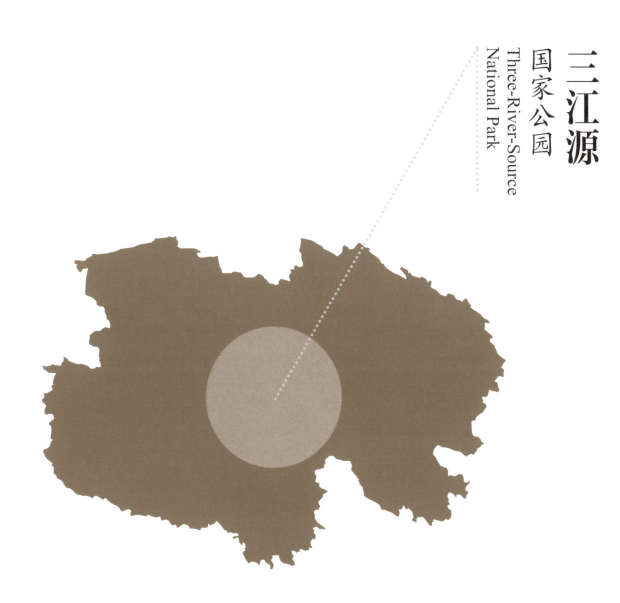

■ 总面积12.31万平方公里

Total Area 123100 km^2

藏羚羊
Pantholops hodgsonii

野牦牛
Bos mutus

Three-River-Source National Park 三江源国家公园

棕熊
Ursus arctos pruinosus

大天鹅
Cygnus cygnus

野驴
Equus kiang

大鵟
Buteo hemilasius

赤狐
Vulpes vulpes

高原鼠兔
Ochotona curzoniae

角百灵
Eremophila alpestris

2016年12月，开始东北虎豹国家公园体制试点。试点区位于吉林、黑龙江两省交界的老爷岭南部区域，总面积1.46万平方公里，其中吉林省片区约占69%，黑龙江省片区约占31%，试点区以中低山、峡谷和丘陵地貌为主，森林面积广阔，森林类型以针阔混交林为主，原生性红松阔叶混交林仅呈零星分布，次生林分布广泛，以白桦林、山杨林、栎林为主。富饶的温带森林生态系统，养育着和庇护着完整的野生动植物群系，是我国东北虎、东北豹最重要的定居和繁育区域，也是重要的温带野生动植物分布区，属于北半球温带区生物多样性最丰富的地区之一。

    In December 2016, the Pilot project of Northeast China Tiger and Leopard National Park was initiated. The Pilot area is located in the southern part of Laoyeling at the junction of Jilin and Heilongjiang provinces, covering a total area of 14600 km^2, of which Jilin Province accounts for about 69% and Heilongjiang Province 31%. The Pilot area is mainly composed of terrains of low-middle mountains, canyons and hills. The vast forest area covers such dominant forest types as mixed coniferous and broad-leaved forest, the sporadic distribution of mixed broad-leaved forest of the primary Korean pine, and such widely-distributed secondary forest as birch, poplar and oak. The rich temperate forest ecosystem, which nurtures and protects the complete wildlife community, is not only the most important settlement and breeding area of Northeast China tiger and leopard in China but also a key distribution area of temperate wildlife. It has become one of the regions with the richest biodiversity in the temperate zone of the Northern Hemisphere.

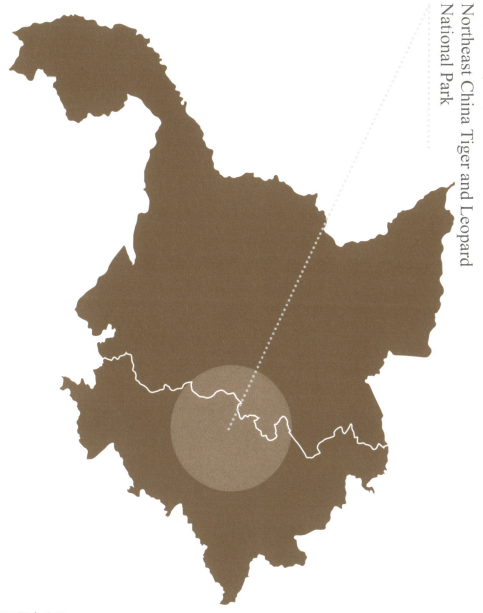

东北虎豹国家公园
Northeast China Tiger and Leopard National Park

- 总面积1.46万平方公里
  Total Area 14600 km²

- 吉林省片区约占69%
  The Pilot area of Jilin Province accounts for about 69%.

- 黑龙江省片区约占31%
  The Pilot area of Heilongjiang Province accounts for about 31%.

东北虎
Panthera tigris altaica

东北虎
Panthera tigris altaica

东北虎
Panthera tigris altaica

东北豹
Panthera pardus orientalis

2016年12月，开始大熊猫国家公园体制试点。大熊猫是我国独有的珍稀物种，也是全球关注的明星物种，具有全球意义的保护价值和极为重要的伞护作用。试点区总面积2.71万平方公里，横跨四川、陕西、甘肃三省12个市（州）30个县（市、区），其中四川省片区约占74.36%，陕西省片区约占16.16%，甘肃省片区约占9.48%。整合各类自然保护地80余个。试点区内有野生大熊猫1631只，国家重点保护野生动物116种，国家重点保护野生植物35种，是全球生物多样性保护热点地区，也是我国生态安全战略格局"两屏三带"的重要区域。

In December 2016, the Pilot project of the Giant Panda National Park was launched. Giant Panda is not only a rare and endemic species in China but also a flagship species of global concern. It has a significant value of global conservation and is playing a leading role in umbrella protection. The Pilot area covers a total area of 27100 km^2, involving 12 cities /prefectures and 30 counties /county-level cities and districts in Sichuan, Shaanxi and Gansu provinces, of which Sichuan Province accounts for about 74.36%, Shaanxi Province 16.16%, and Gansu Province about 9.48%. More than 80 protected areas have been integrated. There are 1631 giant pandas in the wild, 116 species of wild fauna and 35 species of wild flora under national protection in the Pilot area, which is not only a hot spot of global biodiversity conservation but also an important part of "two shelters and three belts", China's strategic pattern of ecological security.

大熊猫国家公园
Giant Panda National Park

- 总面积2.71万平方公里
  Total Area 27100 km^2

- 四川省片区约占74.36%
  The Pilot area of Sichuan Province accounts for about 74.36%.

- 陕西省片区约占16.16%
  The Pilot area of Shaanxi Province accounts for about 16.16%.

- 甘肃省片区约占9.48%
  The Pilot area of Gansu Province accounts for about 9.48%.

大熊猫
Ailuropoda melanoleuca

大熊猫
Ailuropoda melanoleuca

大熊猫
Ailuropoda melanoleuca

小熊猫

Ailurus fulgens

中华斑羚
Naemorhedus griseus

羚牛
Budorcas taxicolor

川金丝猴
Rhinopithecus roxellanae

黄喉貂
Martes flavigula

林麝
Moschus berezovskii

喜马拉雅白眉朱雀
Carpodacus thura

朱鹮
Nipponia nippon

独叶草
Kingdonia uniflora

红豆杉
Taxus wallichiana

珙桐
Davidia involucrata

2017年6月，开始祁连山国家公园体制试点。试点区位于甘肃、青海两省交界，我国西北部、青藏高原东北部，总面积5.02万平方公里，其中甘肃省片区3.44万平方公里，占总面积的68.5%，青海省片区1.58万平方公里，占总面积的31.5%。祁连山是我国西部重要生态安全屏障，是黄河流域重要水源产流地，是我国生物多样性保护优先区域，是冰川与水源涵养类型的国家重点生态功能区之一，国家公园范围内分布的国家一级重点保护野生动物就有15种，如雪豹、白唇鹿、黑鹳、马麝、黑颈鹤等，试点区分布有山地森林、温带荒漠草原、高寒草甸和冰川雪山等复合生态系统。

In June 2017, the Pilot project of the Qilian Mountain National Park was initiated. The Pilot area is located at the junction of Gansu and Qinghai provinces, northwest China and the northeast of the Qinghai-Tibet Plateau, covering a total area of 50200 km^2, including 34400 km^2 in Gansu Province, accounting for 68.5% of the total area, and 15800 km^2 in Qinghai Province, taking up 31.5% of the total area.

Qilian Mountain is an important ecological security barrier in western China, an important source of runoff generation in the Yellow River Basin, a priority area for biodiversity conservation in China, and one of the key national ecological functional areas of glaciers and water conservation types. There are 15 species of wild fauna under Class I national protection distributed in the national park, such as snow leopard, white-lipped deer, black stork, Moschus chrysogaster and black-necked crane. There are compound ecosystems in the Pilot area, such as mountain forest, temperate desert steppe, alpine meadow, glacier and snow mountain.

祁连山国家公园
Qilian Mountain National Park

- 总面积5.02万平方公里
  Total Area 50200 km²

- 甘肃省片区约占68.5%
  The Pilot area of Gansu Province accounts for about 68.5%.

- 青海省片区约占31.5%
  The Pilot area of Qinghai Province accounts for about 31.5%.

Qilian Mountain National Park　祁连山国家公园

雪豹
Panthera uncia

藏狐
Vulpes ferrilata

喜马拉雅旱獭
Marmota himalayana

黑颈鹤
Grus nigricollis

高山兀鹫
Gyps himalayensis

Qilian Mountain National Park 祁连山国家公园

白唇鹿
Cervus albirostris

2019年1月，开始海南热带雨林国家公园体制试点。试点区位于海南岛中部山区，东起吊罗山国家森林公园，西至尖峰岭国家级自然保护区，南自保亭县毛感乡，北至黎母山省级自然保护区，总面积4403平方公里，约占海南岛陆域面积的1/7。海南热带雨林是世界热带雨林的重要组成部分，是亚洲热带雨林和世界季风常绿阔叶林交错带上唯一的"大陆性岛屿型"热带雨林，是我国分布最集中、保存最完好、连片面积最大的热带雨林，拥有全世界、中国和海南独有的动植物种类及种质基因库，是我国热带生物多样性保护的重要地区。

The Pilot project of the Hainan Tropical Rainforest National Park was initiated in January 2019. The Pilot area is located in the central mountain area of Hainan Island, stretching from Diaoluoshan National Forest Park in the east to Jianfengling National Nature Reserve in the west and from Maogan Township, Baoting County in the south to Limushan Provincial Nature Reserve in the north, covering a total area of more than 4403 km^2, accounting for about one-seventh of the land area of Hainan Island. Hainan tropical rainforest is an important part of the world tropical rainforest and the only "continental island" tropical rainforest in the ecotone of the Asian rainforest and world monsoon evergreen broad-leaved forest. It is the most concentrated, best preserved and largest tropical rainforest in China, with endemic wild fauna and flora species and germplasm banks in Hainan, China and the world at large, and is an important area for tropical biodiversity conservation in China.

海南热带雨林国家公园
Hainan Tropical Rainforest National Park

■ 总面积4403平方公里
Total Area 4403 km²

海南黑冠长臂猿
Nomascus hainanus

坡鹿
Cervus eldii

海南孔雀雉
Polyplectron katsumatae

2016年6月，开始武夷山国家公园体制试点。试点区位于福建省北部，总面积1001.41平方公里，区内包含了自然保护区、风景名胜区、国家森林公园、九曲溪光倒刺鲃国家级水产种质资源保护区4种保护地类型。试点区拥有210.7平方公里未受人为破坏的原生性森林植被，是世界同纬度最完整、最典型、面积最大的中亚热带森林生态系统，被誉为"蛇的王国""鸟的天堂""昆虫世界""世界生物模式标本产地""研究亚洲两栖爬行动物的钥匙"。拥有"碧水丹山"特色的典型丹霞地貌景观和新石器时期古越族人留下的历史文化遗迹，具有被誉为"闽邦邹鲁"的数千年历史文化景观，是闻名国内外的"朱子文化"发源地，是世界红茶和乌龙茶的发源地，也是中国唯一的"茶文化艺术之乡"。

In June 2016, the Pilot project of Wuyishan National Park was launched.

The Pilot area is located in the north of Fujian Province, covering a total area of 1001.41km^2. There are five categories of protected areas in the area, i.e. nature reserves, scenic spots, world cultural and natural heritage sites, national forest parks, and Jiuquxi National Aquatic Germplasm Resources Reserve for *Spinibarbus hollandi*.

The Pilot area boasts 210.7 km^2 of undamaged primary forest vegetation, which is the most complete, typical and largest subtropical forest ecosystem at the same latitude in the world. It is known as "the kingdom of snakes", " paradise of birds", " world of insects", "the world biological specimen producer" as well as "the key to the research on Asian amphibians and reptiles". The Pilot area boasts typical Danxia landform with the characteristics of "lucid water and red landform", the historical and cultural relics left by the ancient Yue people in the Neolithic period as well as thousands of years of history and culture famed as "a flourishing culture and education site in Fujian". It is the birthplace of the world-renowned "Zhu Xi Culture", black tea and Oolong tea, and the only "hometown of tea culture and art" in China.

武夷山国家公园
Wuyishan National Park

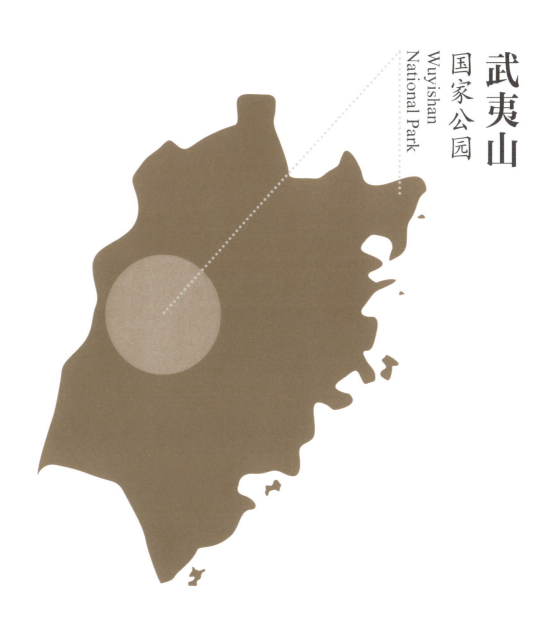

■ 总面积1001.41平方公里

Total Area 1001.41 km²

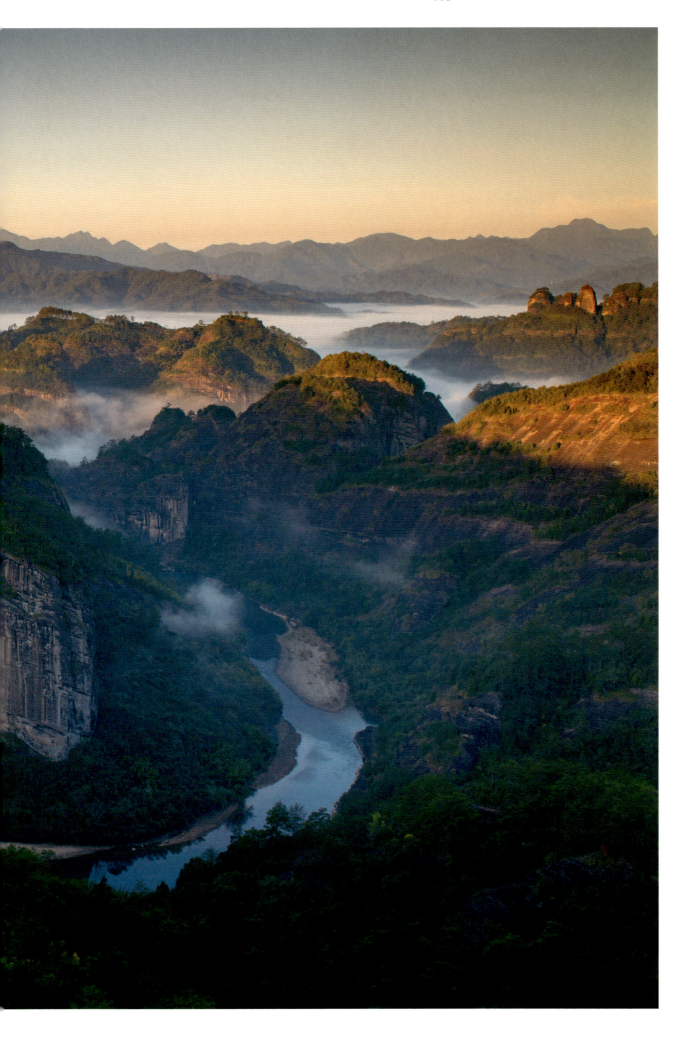

Wuyishan National Park 武夷山国家公园

南方红豆杉
Taxus wallichiana var. mairei

帽蕊草
Mitrastemon yamamotoi

福建假稠李
Maddenia fujianensis

伯乐树（钟萼木）
Bretschneidera sinensis Hemsl.

斑头鸺鹠
Glaucidium cuculoides

黄腹角雉
Tragopan caboti

短尾鸦雀（挂墩鸦雀）
Neosuthora davidiana

红隼
Falco tinnunculus

2016年5月，开始神农架国家公园体制试点。试点区位于湖北省，总面积1170平方公里，是全球性生物多样性王国，拥有被称为"地球之肺"的亚热带森林生态系统、被称为"地球之肾"的泥炭藓湿地生态系统。试点区内分布着中国特有植物及光叶水青冈、洪坪杏等模式植物，以及川金丝猴、林麝等国家一、二级重点保护野生动物，东方白鹳、黑鹳、金雕、白琵鹭、黑冠鹃隼、红腹角雉等国家一、二级重点保护野生鸟类，是中国特有属植物最丰富的地区，是世界生物活化石聚集地和古老、珍稀、特有物种避难所。

In May 2016, the Pilot project of Shennongjia National Park was launched. The Pilot area is located in Hubei Province, covering a total area of 1170 km^2. It is a global kingdom of biodiversity. It boasts a subtropical forest ecosystem known as the "lungs of the earth", a peat moss wetland ecosystem known as the "kidney of the earth" and biodiversity known as the "earth's immune system". In the Pilot area, there are endemic Chinese plants and model plants such as Fagus lucida Rehd. et Wils. and Armeniaca hongpingensis Yü et Li as well as such key wild animals under Class I and II national protection as the golden snub-nosed monkey (Rhinopithecus roxellanae), forest musk deer (Moschus berezovskii), Such wild birds under Class I and II national protection as Oriental white stork (Ciconia boyciana), black stork (Ciconia nigra), golden eagle (Aquila chrysaetos), Eurasian Spoonbill (Platalea leucorodia), Aviceda leuphotes and Tragopan temminckii. It is an area with the most abundant endemic genera in China, a living fossils pool in the world and a refuge for ancient, rare and endemic species.

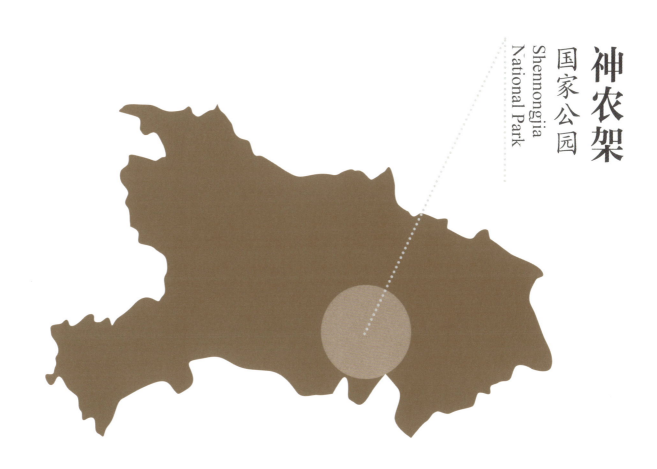

神农架国家公园
Shennongjia National Park

■ 总面积1170平方公里
Total Area 1170 km²

杜鹃
Rhododendron simsii Planch.

冷杉
Abies fabri Mast.Craib

川金丝猴
Rhinopithecus roxellanae

扇脉杓兰
Cypripedium japonicum

七叶一枝花
Paris polyphylla

大鲵
Andrias davidianus

蓝喉太阳鸟
Aethopyga gouldiae

2016年10月，开始普达措国家公园体制试点。试点区位于中国西南横断山脉、云南省迪庆州香格里拉市境内，总面积602.1平方公里，以碧塔海、属都湖和弥里塘亚高山牧场为主要组成部分，园区保持着极为原始状态的寒温性针叶林和硬叶常绿阔叶林，完整的"森林—湖泊—沼泽—草甸"生态系统及丰富的生物多样性，具有很高的保护价值和展示价值。

In October 2016, the Pilot project of Potatso National Park was launched.

The Pilot area is located in the Hengduan Mountains in southwest China and Shangri-La City, Diqing Prefecture, Yunnan Province, covering a total area of 602.1 km^2, with Bitahai Lake, Shudu Lake and Militangya alpine pastures as the main components. The park maintains extremely primitive cold-temperate coniferous forest and hard-leaved evergreen broad-leaved forest, a complete "forest-lake-swamp-meadow" ecosystem and rich biodiversity, boasting high protection value and display value.

普达措国家公园
Potatso National Park

■ 总面积602.1平方公里

Total Area 602.1 km²

Potatso National Park 普达措国家公园

塔黄
Rheum nobile

横断山绿绒蒿
Meconopsis pseudointegrifolia

153   Potatso National Park   普达措国家公园

绿绒蒿
Meconopsis

黑颈鹤
Grus nigricollis

德拜鹦鹉（大紫胸鹦鹉）
Psittacula derbiana

血雉
Ithaginis cruentus

绿尾虹雉
Lophophorus lhuysii

2016年6月，开始钱江源国家公园体制试点。试点区位于浙江省开化县，地处浙皖赣三省交界处，总面积252平方公里。包括古田山国家级自然保护区、钱江源国家森林公园、钱江源省级风景名胜区以及上述自然保护地之间的连接地带。试点区拥有较为完整的低海拔中亚热带常绿阔叶林，是连接华南—华北植物的典型过渡带，保存有大片原始状态的天然次生林。试点区内有黑麂、白颈长尾雉、白鹇等国家重点保护野生动物，是中国特有的世界珍稀濒危物种的主要栖息地之一，是华东地区重要的生态屏障。

In June 2016, the Pilot project of Qianjiangyuan National Park was initiated. The Pilot area is located in Kaihua County, Zhejiang Province, at the junction of Zhejiang, Anhui and Jiangxi provinces, covering a total area of 252 km^2, including Gutianshan National Nature Reserve, Qianjiangyuan National Forest Park, Qianjiangyuan Provincial Scenic Spot and the connection zone between the above-mentioned protected areas. The Pilot area has a comparatively complete low-altitude mid-subtropical evergreen broad-leaved forest, which is a typical transition zone of plants linking South China and North China and preserves a large area of natural secondary forest in the primitive state. There are black muntjac (Muntiacus crinifrons), Elliot's Pheasant (Syrmaticus ellioti), silver pheasant (Lophura nycthemera) and other wild animals under national protection in the Pilot area, which is one of the main habitats of rare and endangered species endemic to China and an important ecological barrier in East China.

钱江源国家公园
Qianjiangyuan National Park

- 总面积252平方公里
- Total Area 252 km²

白鹇
Lophura nycthemera

白颈长尾雉
Syrmaticus ellioti

2016年7月，开始南山国家公园体制试点。试点区位于湖南邵阳城步苗族自治县，主要由原湖南南山国家级风景名胜区、金童山国家自然保护区、两江峡谷国家森林公园、白云湖国家湿地公园4个国家级保护地和部分具有保护价值的区域整合而成，总面积635.94平方公里。试点区地处南岭山系主峰区域，是我国南北纵向山脉交汇枢纽。区内生物多样性极其丰富，有国家重点保护一级野生动物林麝等，一级重点保护野生植物资源冷杉等。

In July 2016, the Pilot project of Nanshan National Park was launched. The Pilot area is located in Chengbu Miao Autonomous County, Shaoyang, Hunan Province. It is mainly composed of 4 national protected areas, i.e. the former Nanshan National Scenic Spot, Jintong Mountain National Nature Reserve, Liangjiang Canyon National Forest Park, Baiyun Lake National Wetland Park and some other area of conservation value, covering a total area of 635.94 km². The Pilot area is located in the main peak area of the Nanling mountain system and at the junction of the north-south longitudinal mountains in China. The area is extremely rich in biodiversity, including such Class I wild fauna under national priority protection as Rumbaugh (Neofelis nebulosa Griffith) and forest musk deer (Moschus berezovskii) and such Class I wild flora resources under national priority protection as fir (Abies fabri (Mast.) Craib).

南山国家公园
Nanshan National Park

■ 总面积635.94平方公里
Total Area 635.94 km^2

中华秋沙鸭
Mergus squamatus

松雀鹰
Accipiter virgatus

鸳鸯
Aix galericulata

Nanshan National Park 南山国家公园

迁徙的鹭群

光叶水青冈
Fajus lucida

资源冷杉
Abies beshanzuensis var. *ziyuanensis*

2013

2020

大事记

2013

11月　党的十八届三中全会首次提出建立国家公园体制。

The Third Plenary Session of the 18th CPC Central Committee proposed a national park system for the first time in November 2013.

2015

1月　国家发展改革委等13部委联合发布了《建立国家公园体制试点方案》。

Thirteen ministries and commissions including the National Development and Reform Commission jointly released the *Pilot Program for a National Park System* in January 2015.

9月　中共中央、国务院印发《生态文明体制改革总体方案》。

The CPC Central Committee and the State Council issued the *General Plan for Reform for Promoting Ecological Progress* in September 2015.

12月　中央全面深化改革领导小组审议通过《三江源国家公园体制试点方案》。

The Central Leading Group for Comprehensively Continuing Reform reviewed and adopted the *Pilot Plan for the Three-River-Source National Park System* in December 2015.

2016

1月　习近平总书记在中央财经领导小组第十二次会议上听取原国家林业局关于森林生态安全工作汇报后，特别指出，要着力建设国家公园，保护自然生态系统的原真性和完整性，给子孙后代留下一些自然遗产。

Following the debriefing of the former State Forestry Administration on forest ecological security at the 12th meeting of the CPC Central Leading Group for Financial and Economic Affairs in January 2016, General Secretary Xi Jinping particularly pointed out that efforts should be made to develop national parks and protect the authenticity and integrity of the natural ecosystem so as to leave some natural heritage to future generations.

3月 中共中央办公厅、国务院办公厅印发《三江源国家公园体制试点方案》。

In March, the General Office of the CPC Central Committee and the General Office of the State Council released the *Pilot Program for the Three-River-Source National Park System*.

5月 国家发展改革委批复《神农架国家公园体制试点区试点实施方案》。

In May, the National Development and Reform Commission approved the Pilot Implementation Plan for the *Pilot project area of Shennongjia National Park System*.

6月 三江源国家公园管理局挂牌成立。

Three-River-Source National Park Service was established in June 2016.

6月 国家发展改革委批复《武夷山国家公园体制试点区试点实施方案》。

In June, the National Development and Reform Commission adopted *the Pilot Implementation Plan for the Pilot project area of Wuyishan National Park System*.

6月 国家发展改革委批复《钱江源国家公园体制试点区试点实施方案》。

In June, the National Development and Reform Commission approved the *Pilot Implementation Plan for the Pilot project area of Qianjiangyuan National Park System*.

7月 国家发展改革委批复《南山国家公园体制试点区试点实施方案》。

In July, the National Development and Reform Commission approved the *Pilot Implementation Plan for the Pilot project area of Nanshan National Park System*.

10月 国家发展改革委批复《香格里拉普达措国家公园体制试点区试点实施方案》。

In October, the National Development and Reform Commission adopted the *Pilot Implementation Plan for the Pilot project area of Shangri-La Potatso National Park System*.

11月	神农架国家公园管理局挂牌成立。 Shennongjia National Park Service was established in November 2016.
12月	中央全面深化改革领导小组审议通过《大熊猫国家公园体制试点方案》《东北虎豹国家公园体制试点方案》。 The Central Leading Group for Comprehensively Continuing Reform examined and approved the *Pilot Program for the Giant Panda National Park System* and the *Pilot Program for Northeast China Tiger and Leopard National Park System* in December 2016.

2017

1月	中共中央办公厅、国务院办公厅印发《东北虎豹国家公园体制试点方案》《大熊猫国家公园体制试点方案》。 In January, the General Office of the CPC Central Committee and the General Office of the State Council released the *Pilot Program for Northeast China Tiger and Leopard National Park System* and the *Pilot Program for the Giant Panda National Park System*.
6月	中央全面深化改革领导小组审议通过《祁连山国家公园体制试点方案》。 The Central Leading Group for Comprehensively Continuing Reform examined and approved the *Pilot Program for the Qilian Mountain National Park System* in June 2017.
6月	武夷山国家公园管理局正式成立。 Wuyishan National Park Service was officially established in June 2017.
7月	湖北省政府批复《神农架国家公园总体规划》。 The Hubei Provincial Government adopted the *Master Plan of the Shennongjia National Park* in July 2017.
8月	东北虎豹国家公园管理局挂牌成立。 Northeast China Tiger and Leopard National Park Service was established in August 2017.
8月	《三江源国家公园条例（试行）》施行。 The *Regulations on Three-River-Source National Park (for trial implementation)* was implemented as of August 2017.

9月　中共中央办公厅、国务院办公厅印发《建立国家公园体制总体方案》。

The General Office of the CPC Central Committee and the General Office of the State Council issued the *General Plan for the Establishment of a National Park System* in September 2017.

9月　中共中央办公厅、国务院办公厅印发《祁连山国家公园体制试点方案》。

In September, the General Office of the CPC Central Committee and the General Office of the State Council released the *Pilot Program for the Qilian Mountain National Park System*.

10月　党的十九大报告指出，建立以国家公园为主体的自然保护地体系。

The report of the 19th CPC National Congress pointed out in October 2017 that a national park-centered protected area system should be established.

10月　南山国家公园管理局挂牌成立。

Nanshan National Park Service was established in October 2017.

10月　浙江省发展改革委批复《钱江源国家公园体制试点区总体规划（2016-2025）》。

The Development and Reform Commission of Zhejiang Province approved the *General Plan of the Pilot Area of Qianjiangyuan National Park System (2016-2025)* in October 2017.

2018

1月　经国务院同意，国家发展改革委印发《三江源国家公园总体规划》。

With the consent of the State Council, the National Development and Reform Commission issued the *Master Plan of Three-River-Source National Park* in January 2018.

2月　国家林业局东北虎豹监测与研究中心在北京师范大学揭牌成立。

The Research and Monitoring Center for Northeast China Tiger and Leopard, the State Forestry Administration was unveiled in Beijing Normal University in February 2018.

3月　《武夷山国家公园条例（试行）》施行。
The Regulations on (for trial implementation) Wuyishan National Park was implemented as of March 2018.

4月　国家林业和草原局（国家公园管理局）正式挂牌。
The National Forestry and Grassland Administration (National Park Administration) was unveiled in April 2018.

5月　《神农架国家公园保护条例》施行。
The Regulations on Protection of the Shennongjia National Park was implemented as of May 2018.

8月　普达措国家公园管理局挂牌成立。
Potatso National Park Service was established in August 2018.

10月　祁连山国家公园管理局和大熊猫国家公园管理局挂牌成立。
Qilian Mountain National Park Service and the Giant Panda National Park Service were established in October 2018.

11月　清华大学国家公园研究院挂牌成立。
The Research Institute of National Parks, Tsinghua University was established in November 2018.

12月　国家林业和草原局国家公园规划研究中心在国家林业和草原局昆明勘察设计院挂牌成立。
The National Park Planning and Research Center, the National Forestry and Grassland Administration was launched in the Kunming Inventory and Design Institute, the National Forestry and Grassland Administration in December 2018.

2019

1月　中央全面深化改革委员会第六次会议审议通过《关于建立以国家公园为主体的自然保护地体系的指导意见》《海南热带雨林国家公园体制试点方案》。
The sixth meeting of The Central Leading Group for Comprehensively Continuing Reform examined and approved the *Guidelines on Establishment of a National Park-Centered Protected Area System* and the *Pilot Plan for Hainan Tropical Rainforest National Park System* in January 2019.

4月　海南热带雨林国家公园管理局挂牌成立。

Hainan Tropical Rainforest National Park Service was established in April 2019.

4月　国家林业和草原局国家公园监测评估研究中心在国家林业和草原局调查规划设计院挂牌成立。

The National Park Monitoring, Evaluation and Research Center, the National Forestry and Grassland Administration was established in the Inventory, Planning and Design Institute of the National Forestry and Grassland Administration in April 2019.

6月　中共中央办公厅、国务院办公厅印发《关于建立以国家公园为主体的自然保护地体系的指导意见》。

The General Office of the CPC Central Committee and the General Office of the State Council issued the *Guidelines on the Establishment of a National Park-Centered Protected Area System* in June 2019.

7月　国家公园管理局印发《海南热带雨林国家公园体制试点方案》。

The National Park Administration released the *Pilot Program of Hainan Tropical Rainforest National Park System* in July 2019.

7月　钱江源国家公园管理局挂牌成立。

Qianjiangyuan National Park Service was established in July 2019.

7月　国家公园管理局办公室印发《国家公园管理局办公室关于开展国家公园体制试点评估工作的函》，全面启动国家公园体制试点第三方评估工作。

The General Office of the National Park Administration issued a *Letter of the Office of the National Park Administration on Carrying Out the Evaluation of the Pilot Program of the National Park System* in July 2019, launching the third-party evaluation of the Pilot program of the national park system.

8月　国家林业和草原局（国家公园管理局）与青海省人民政府举办"第一届国家公园论坛"，习近平总书记发来贺信。

The first National Park Forum was held in August 2019 by the National Forestry and Grassland Administration (National Park Administration) and the People's Government of Qinghai Province. General Secretary Xi Jinping sent a congratulatory letter.

10月	党的十九届四中全会提出"要加强对重要生态系统的保护和永续利用，构建以国家公园为主体的自然保护地体系，健全国家公园保护制度"。
	The Fourth Plenary Session of the 19th CPC Central Committee proposed to strengthen the protection and sustainable use of key ecosystems, build a national park-centered protected area system, and improve the national park protection system in October 2019.
10月	国家林业和草原局（国家公园管理局）发布公告，决定成立国家公园和自然保护地标准化技术委员会。
	The National Forestry and Grassland Administration (National Park Administration) issued a notice to establish a Technical Committee for Standardization of National Parks and Protected Areas in October 2019.
11月	福建省委通过《武夷山国家公园总体规划》。
	The Fujian Provincial Party Committee approved the *Master Plan of Wuyishan National Park* in November 2019.
12月	国家林业和草原局（国家公园管理局）召开国家公园体制试点工作会议。
	The National Forestry and Grassland Administration (National Park Administration) hosted a work meeting on the Pilot program of the national park system in December 2019.
12月	国家林业和草原局（国家公园管理局）全面支持青海推进以国家公园为主体的自然保护地体系示范省建设。
	The National Forestry and Grassland Administration (National Park Administration) offered full support to Qinghai in promoting the development of a demonstration province for a national park-centered protected area system in December 2019.

2020

1月	东北虎豹国家公园"天地空"一体化监测体系中试开通，覆盖面积达5000平方公里。
	The intermediate test of the 3-D integrated monitoring system for Northeast China Tiger and Leopard National Park was launched in January 2020, covering an area of 5000 km^2.

3月　国家林业和草原局（国家公园管理局）批准发布《国家公园总体规划技术规范》（LY/T 3188-2020）《国家公园资源调查与评价规范》（LY/T 3189-2020）《国家公园勘界立标规范》（LY/T 3190-2020）3个林业行业标准。

The National Forestry and Grassland Administration (National Park Administration) approved and issued three forest industry standards in March 2020, i.e. *Technical Code for General Plan of National Parks* (LY/T 3188-2020), *Resources Inventory and Evaluation Code for National Parks* (LY/T 3189-2020) and *Demarcation Code for National Parks* (LY/T 3190-2020).

4月　国家林业和草原局国家公园建设咨询研究中心在国家林草局林产工业规划设计院挂牌成立。

The Consulting and Research Center for Development of National Parks, the National Forestry and Grassland Administration was unveiled in the Planning and Design Institute of Forest Products Industry, the National Forestry and Grassland Administration in April 2020.

4月　云南省人民政府通过《香格里拉普达措国家公园总体规划》。

The People's Government of Yunnan Province adopted the *Master Plan of Shangri-La Potatso National Park* in April 2020.

4月　湖南省邵阳市人民政府通过《南山国家公园总体规划（2018—2025年）》；湖南省邵阳市人民政府办公室印发《南山国家公园管理办法》。

The People's Government of Shaoyang City, Hunan Province adopted the *Master Plan of Nanshan National Park (2018-2025)*, and the Office of the People's Government of Shaoyang City, Hunan Province released the *Regulations on the Management of Nanshan National Park* in April 2020.

6月　国家林业和草原局（国家公园管理局）全面启动国家公园体制试点第三方评估验收工作。

The National Forestry and Grassland Administration (National Park Administration) launched the third-party evaluation and acceptance of the Pilot program of the national park system in June 2020.

7月　国家林业和草原局（国家公园管理局）印发《东北虎豹国家公园总体规划》《祁连山国家公园总体规划》《大熊猫国家公园总体规划》和《海南热带雨林国家公园总体规划》。

The National Forestry and Grassland Administration (National Park Administration) released in July 2020 the *Master Plan of Northeast China Tiger and Leopard National Park*, the *Master Plan of the Qilian Mountain National Park*, the *Master Plan of the Giant Panda National Park* as well as the *Master Plan of Hainan Tropical Rainforest National Park*.

8月　国家林业和草原局（国家公园管理局）召开国家公园建设座谈会。

The National Forestry and Grassland Administration (National Park Administration) convened a workshop on national park development in August 2020.

结语

CONCLUDING
REMARKS

一次次会议的召开，一份份文件的出台，一个个管理机构的成立，这些都标志着国家公园体制试点工作有序展开，稳步前行。到2020年，完成国家公园体制试点，设立一批国家公园。到2025年，健全国家公园体制，初步建成以国家公园为主体的自然保护地体系。到2035年，自然保护地规模和管理达到世界先进水平，全面建成中国特色自然保护地体系。

但是，如何推动已有各类自然保护地的分类、定位、整合；如何处理好以国家公园为主体的自然保护地的保护与发展关系；如何在保护的基础上发挥国家公园科研、教育、游憩、社区发展等功能，仍然面临巨大的挑战，任重而道远……

长风破浪会有时，直挂云帆济沧海。作为具有独特价值，并具有国家和全球意义的中国国家公园，与每一位国人息息相关。我们将肩负重任，不懈努力建设美丽中国，为治理全球生态提供中国智慧、中国经验和中国方案。

中国的国家公园，必将迎来更为辉煌的未来！

The convening of meetings, the release of documents and the establishment of management agencies indicate that the Pilot trial of the national park system has been progressing in a stepwise manner. By 2020, the Pilot trial of the national park system will be completed with a number of national parks established. By 2025, the national park system will be improved and a national park-centered protected area system will take shape. By 2035, the scale and management of protected areas will reach the world's advanced level, and a protected area system with Chinese characteristics will be established in an all-round way.

However, such issues as how to promote the classification, positioning, and integration of all kinds of protected areas, how to deal with the relationship between protection and development of protected areas represented by national parks and how to give play to the scientific research, education, recreation, community development and other functions of national parks based on protection are still facing great challenges and have a long way to go.

One day, we will be in the waves at full speed to show our aspirations; hang our sailcloth up, going across the sea, we will achieve our dream. Chinese national parks boast endemic value and national and global significance. They are closely related to every countryman. We will shoulder the important mission of making unremitting efforts to build a beautiful China and contribute to global ecological governance by offering the Chinese wisdom, experiences and solutions. China's national parks will usher in a more brilliant future.

图书在版编目（CIP）数据

中国国家公园.试点篇/国家林业和草原局编.--

北京：中国林业出版社，2020.12

ISBN 978-7-5219-0848-0

Ⅰ.①中… Ⅱ.①国… Ⅲ.①国家公园—建设—中国

Ⅳ.① S759.992

中国版本图书馆CIP数据核字(2020)第197627号

中国林业出版社·国家公园分社·文化产业事业部

策 划 人：刘东黎

责任编辑：张衍辉　葛宝庆　黄晓飞

书籍设计：芥子设计

设计协力：罗　兵

电　　话：（010）83143521　83143612

出版发行：中国林业出版社

（100009，北京市西城区刘海胡同7号）

E-mail：np83143521@126.com

经　　销：新华书店

印　　刷：北京雅昌艺术印刷有限公司

版　　次：2020年12月第1版

印　　次：2020年12月第1次印刷

开　　本：889mm×1194mm　1/8

印　　张：18.75

字　　数：288千字

定　　价：268.00元

审图号：GS（2021）63号

未经许可，不得以任何方式复制或抄袭本书之部分或全部内容，版权所有，侵权必究。